Emma Parfitt and James Petch

Zib Zab's Space Adventure

Emma Parfitt

Illustrated by James Petch

An adventure with Zib Zab is never far away,

So off we all go to visit the **Milky Way**.

Sat in the spaceship,
he is ready for the flight,

3 . . 2 . . 1 . .
LIFT OFF !!

And he whizzes out of sight.

Off into the darkness Zib Zab travels far
Until he finds the **Sun**, our large, hot star.

Rotating around the **Sun** all the planets go.
But how many planets? Do you know?

Zib Zab cannot land as he doesn't want to burn,
So off to **Mercury** he makes the spaceship turn.

Travelling towards it, this planet number **one**,
Zib Zab discovers it's the closest to the **Sun**.

Mercury is really hot and also rather small.
It's just like a great big, rocky, iron ball.

3..2..1..Blast off! He soars into the night,
Heading towards **Venus** shining really bright.

Beneath its poisonous clouds **Venus** is hiding,
A hot and rocky planet Zib Zab is finding.

Venus is a planet which spins the other way
And it's not a very nice place for us to stay.

So up to his spaceship he did climb
To find another planet, yes it's time!

Next is **Earth,** planet number **three.**
This is where you live! Can you see?

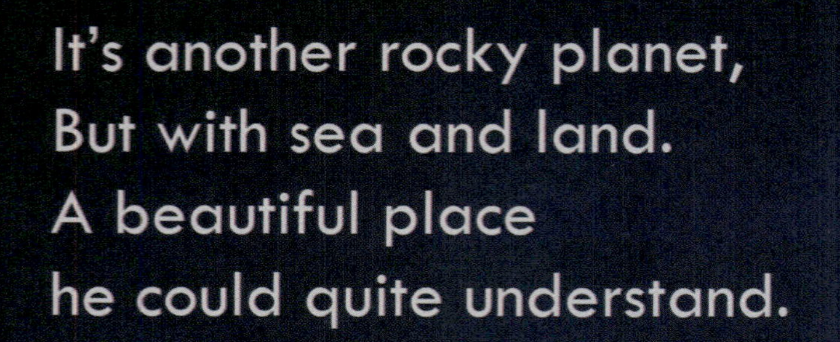

It's another rocky planet,
But with sea and land.
A beautiful place
he could quite understand.

Earth's **Moon** in the sky is shining bright,
So let's explore there on our next flight.

So off he whizzes to reach the **Moon**
And hopes he'll be there very soon.

The **Moon** goes around **Earth** all the time,
As out of his spaceship Zib Zab did climb.

Zib Zab spies footprints on the **Moon's** floor,
Meaning spacemen have been here before.

On the surface these footprints will stay,
As there is no wind to blow them away.

From the **Moon** another planet he does see
And Zib Zab wonders which one it could be.

So up to his spaceship he did climb
To find another planet, yes it's time!

Zooming through space and the sparkly stars,
Zib Zab arrives on the red planet **Mars.**

He realises that this is planet number **four**
As he gets out and closes his spaceship door.

The mountains on **Mars** are extremely high
And there's no water, so it's really dry.

Do martians really live here? Nobody knows,
He would like to find some friends,

But off he goes . . .

Up to his spaceship he does climb,
To find another planet, yes it's time!

Between **Mars** and **Jupiter** the **Asteroid Belt** lies,

Big floating rocks in the dark starry skies.

Zib Zab's spaceship knows it should avoid

Being hit by a great **big** asteroid!

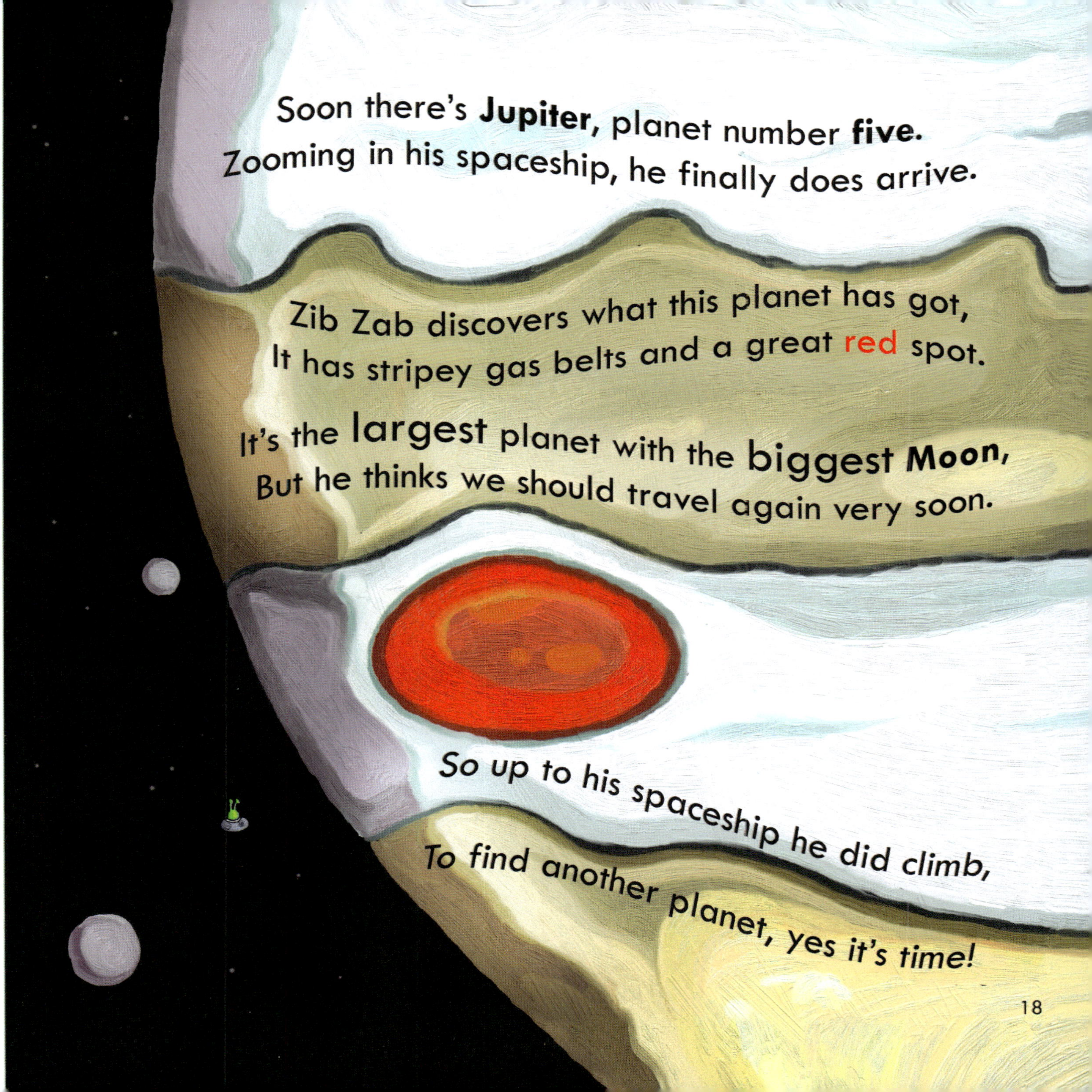

Soon there's **Jupiter**, planet number **five.**
Zooming in his spaceship, he finally does arrive.

Zib Zab discovers what this planet has got,
It has stripey gas belts and a great red spot.

It's the largest planet with the biggest **Moon**,
But he thinks we should travel again very soon.

So up to his spaceship he did climb,
To find another planet, yes it's time!

Zooming really fast Zib Zab was quite keen.
The next stop was **Saturn,** the best he had seen!
This planet is an amazing sight, it really does entice.

Number **six** is **Saturn** with its rings of dust and ice.
Saturn is a gas planet in our **Solar System,**
With lots of Moons, you cannot miss them.

On **Saturn** Zib Zab would like to stay
But he wants to find more planets on which to play!

So up to his spaceship he did climb
To find another planet, yes it's time!

The first by telescope **Uranus** was found,
As Zib Zab continued to explore around.

Uranus, the coldest planet as it's far away.
Spinning on its side, it tilts the wrong way!

This planet number **seven** has thin **black** rings,
But it's time to see what the next one brings.

In the distance a big blue planet he has seen,
And plans to leave here and its clouds of green.

So up to his spaceship
he did climb,
To find another planet,
yes it's time!

Next is planet number **eight** where it's very dark,
As Zib Zab struggles to see somewhere to park!

This place is blue and icy, it's never very warm.
Extremely cold and windy, there's always a storm.

The planet **Neptune** is the furthest from the **Sun**
Our journey's nearly over, we've had such fun!

So back out into space Zib Zab did blast,
On a special spacecraft zooming very fast!

So Zib Zab leaves
the gas giants behind,
As a small dwarf planet
he plans to find.

*To the edge of the **Solar System** Zib Zab flies,*
*Where a ball of rock and ice called **Pluto** lies.*

Upon **Pluto** Zib Zab would like to roam
But now it's time to travel back home.

Did you enjoy the tour of our **Solar System**?
With so many planets, did you count them?

"Would you like to come with me and explore?
We could have fun finding so much more!"

So up to his spaceship he did climb,
To travel into deep space, yes it's time!

Illustrated Constellations

Inside Cover: **The constellation of Hercules**
Hercules, the Roman mythological hero. Best visible in July.

Page 5: **The constellation of Orion**
Orion, a hunter in Greek mythology. Best visible in January.

Page 6: **The constellation of Cassiopeia**
Cassiopeia, a vain queen in Greek mythology. Best visible in November.

Page 7: **The constellations of Andromeda and Pegasus**
Andromeda, daughter of Cassiopeia in Greek mythology. Best visible in November.
Pegasus, a winged horse in Greek mythology. Best visible in October.

Page 8: **The constellation of Leo**
Lion killed by the mythical Greek hero Heracles (Hercules). Best visible in April.

Page 10: **The constellations of Delphinus and Sagitta**
Delphinus, the Latin name for dolphin. Best visible in September.
Sagitta, the Latin name for arrow. Best visible in August.

Page 12: **The constellations of Cygnus and Lyra**
Cygnus, the Latin name for swan. Best visible in September.
Lyra, the lyre (harp) of Orpheus in Greek mythology. Best visible in August.

Page 16: **The constellations of Ursa Major and Ursa Minor**
Ursa Major, the Latin name for great bear. Best visible in April.
Ursa Minor, the latin name for little bear. Best visible in June.

Page 20: **The constellations of Cancer and Gemini**
Cancer, the Latin name for crab. Best visible in March.
Gemini, the Latin name for twins. Best visible n February.

Page 21: **The constellation of Taurus**
Taurus, the Latin name for bull. Best visible in January.

Page 23: **The constellation of Perseus**
Perseus, the Greek mythological hero. Best visible in December.

Page 24. **The constellation of Pisces**
Pisces, the Latin name for fish. Best visible in November.

Page 25: **The constellation of Sagittarius**
Sagittarius, the Latin name for archer. Best visible in August.

Page 26: **The constellation of Serpens**
Serpens, a snake held by the healer Asclepius in Greek mythology. Best visible in July.

Page 27: **The constellations of Corona Borealis and Bootes**
Corona Borealis, the Latin name for northern crown. Best visible in July.
Bootes, a herdsman in Greek mythology. Best visible in June.

Printed in Great Britain
by Amazon